Man a Machine

Julien Offray de La Mettrie

1

It is not enough for a wise man to study nature and truth; he should dare state truth for the benefit of the few who are willing and able to think. As for the rest, who are voluntarily slaves of prejudice, they can no more attain truth, than frogs can fly.

I reduce to two the systems of philosophy which deal with man's soul. The first and older system is materialism; the second is spiritualism.

The metaphysicians who have hinted that matter may well be endowed with the faculty of thought have perhaps not reasoned ill. For there is in this case a certain advantage in their inadequate way of expressing their meaning. In truth, to ask whether matter can think, without considering it otherwise than in itself, is like asking whether matter can tell time. It may be foreseen that we shall avoid this reef upon which Locke had the bad luck to shipwreck.

The Leibnizians with their monads have set up an unintelligible hypothesis. They have rather spiritualized matter than materialized the soul. How can we define a being whose nature is absolutely unknown to us?

Descartes and all the Cartesians, among whom the followers of Malebranche have long been numbered, have made the same mistake. They have taken for granted two distinct substances in man, as if they had seen them, and positively counted them.

The wisest men have declared that the soul can not know itself save by the light of faith. However, as reasonable beings they have thought that they could reserve for themselves the right of examining what the Bible means by the word "spirit," which it uses in speaking of the human soul. And if in their investigation, they do not agree with the theologians on this point, are the theologians more in agreement among themselves on all other points?

Here is the result in a few words of all their reflections. If there is a God, he is the Author of nature was well as of revelation. He has given us the one to explain the other, and reason to make them agree.

To distrust the knowledge that can be drawn from the study of animated bodies, is to regard nature and revelation as two contraries which destroy each other, and consequently to dare uphold the absurd doctrine, that God contradicts Himself in His various works and deceives us.

If there is a revelation, it can not then contradict nature. By nature only can we understand the meaning of the words of the Gospel, of which experience is the only truly interpreter. In fact, the commentators before our time have only obscured the truth. We can judged of this by the author of the Spectacle of Nature. "It is astonishing," he says concerning Locke, "that a man who degrades our soul far enough to consider it a soul of clay should dare set up reason as judge and sovereign arbiter of the mysteries of faith, for," he adds, "what an astonishing idea of Christianity one would have, if one were to follow reason."

Not only do these reflections fail to elucidate faith, but they also constitute such frivolous objections to the method of those who undertake to interpret the

Scripture, that I am almost ashamed to waste time in refuting them.

The excellence of reason does not depend on a big word devoid of meaning (immateriality), but on the force, extent, and perspicuity of reason itself. Thus a "soul of clay" which should discover, at one glance, as it were, the relations and the consequences of an infinite number of ideas hard to understand, would evidently be preferable to a foolish and stupid soul, though that were composed of the most precious elements. A man is not a philosopher because, with Pliny, he blushes over the wretchedness of our origin. What seems vile is here the most precious of things, and seems to be the object of nature's highest art and most elaborate care. But as man, even though he should come from an apparently still more lowly source, would yet be the most perfect of all beings, so whatever the origin of his soul, if it is pure, noble, and lofty, it is a beautiful soul which dignifies the man endowed with it.

Pluche's second way of reasoning seems vicious to me, even in his system, which smacks a little of fanaticism; for [on his view] if we have an idea of faith as being contrary to the clearest principles, to the most incontestable truths, we must yet conclude, out of respect for revelation and its author, that this conception is false, and that we do not yet understand the meaning of the words of the Gospel.

Of the two alternatives, only one is possible: either everything is illusion, nature as well as revelation, or experience alone can explain faith. But what can be more ridiculous than the position of our author! Can one imagine hearing a Peripatetic say, "We ought not to accept the experiments of Torricelli, for if we should accept them, if we should rid ourselves of the horror of the void, what an astonishing philosophy we should have!"

I have shown how vicious the reasoning of Pluche is in order to prove, in the first place, that if there is a revelation, it is not sufficiently demonstrated by the mere authority of the Church, and without any appeal to reason, as all those who fear reason claim: and in the second place, to protect against all assault the method of those who would wish to follow the path that I open to them, of interpreting supernatural things, incomprehensible in themselves, in the light of those ideas with which nature has endowed us. Experience and observation should therefore be our only guides here. Both are to be found throughout the records of the physicians who were philosophers, and not in the works of the philosophers who were not physicians. The former have traveled through and illuminated the labyrinth of man; they alone have laid bare those springs [of life] hidden under the external integument which conceals so many wonders from our eyes. They alone, tranquilly contemplating our soul, have surprised it, a thousand times, both in its wretchedness and in its glory, and they have no more despised it in the first estate, than they have admired it in the second. Thus, to repeat, only the physicians have a right to speak on this subject. What could the others, especially the theologians, have to say? Is it not ridiculous to hear them shamelessly coming to conclusions about a subject concerning which they have had no means of knowing anything, and from which on the contrary they have been completely turned aside by obscure studies that have led them to a thousand prejudiced opinions, - in a word, to fanaticism, which adds yet more to their ignorance of the mechanism of the body?

But even though we have chosen the best guides, we shall still find many thorns and stumbling blocks in the way.

Man is so complicated a machine that it is impossible to get a clear idea of the machine beforehand, and hence impossible to define it. For this reason, all the investigations have been vain, which the greatest philosophers have made à priori, that is to to say, in so far as they use, as it were, the wings of the spirit. Thus it is only à posteriori or by trying to disentangle the soul from the organs of the body, so to speak, that one can reach the highest probability concerning man's own nature, even though one can not discover with certainty what his nature is.

Let us then take in our hands the staff of experience, paying no heed to the accounts of all the idle theories of the philosophers. TO be blind and to think one can do without this staff if the worst kind of blindness. How truly a contemporary writer says that the only vanity fails to gather from secondary causes the same lessons as from primary causes! One can and one even ought to admire all these fine geniuses in their most useless works, such men as Descartes, Malebranche, Leibnitz, Wolff and the rest, but what profit, I ask, has any one gained from their profound meditations, and from all their works? Let us start out then to discover not what has been thought, but what must be thought for the sake of repose in life.

There are as many different minds, different characters, and different customs, as there are different temperaments. Even Galen knew this truth which Descartes carried so far as to claim that medicine alone can change minds and morals, along with bodies. (By the write of L'historie de l'âme, this teaching is incorrectly attributed to Hippocrates.) It is true that melancholy, bile, phlegm, blood etc., - according to the nature, the abundance, and the different combination of these humors - make each man different from another.

In disease the soul is sometimes hidden, showing no sign of life; sometimes it is so inflamed by fury that it seems to be doubled; sometimes, imbecility vanishes and the convalescence of an idiot produces a wise man. Sometimes, again, the greatest genius becomes imbecile and looses the sense of self. Adieu then to all that fine knowledge, acquired at so high a price, and with so much trouble! Here is a paralytic who asks is his leg is in bed with him; there is a soldier who thinks that he still has the arm which has been cut off. The memory of his old sensations, and of the place to which they were referred by his soul, is the cause of this illusion, and of this kind of delirium. The mere mention of the member which he has lost is enough to recall it to his mind, and to make him feel all its motions; and this causes him an indefinable and inexpressible kind of imaginary suffering. This man cries like a child at death's approach, while this other jests. What was needed to change the bravery of Caius Julius, Seneca, or Petronius into cowardice or faintheartedness? Merely an obstruction in the spleen, in the liver, an impediment in the portal vein. Why? Because the imagination is obstructed along with the viscera, and this gives rise to all the singular phenomena of hysteria and hypochondria.

What can I add to the stories already told of those who imagine themselves transformed into wolf-men, cocks or vampires, or of those who think that the dead feed upon them? Why should I stop to speak of the man who imagines that his nose or some other member is of glass? The way to help this man to regain his faculties and his own flesh-and-blood nose is to advise him to sleep on hay, lest he beak the fragile organ, and then to set fire to the hay that he may be afraid of being burned - a far which has sometimes cured paralysis. But I must touch lightly on facts which everybody knows.

Neither shall I dwell long on the details of the effects of sleep. Here a tired soldier snores in a trench, in the middle of the thunder of hundreds of cannon. His soul hears nothing; his sleep is as deep as apoplexy. A bomb is on the point of crushing him. He will feel this less perhaps than he feels an insect which is under his foot.

On the other hand, this man who is devoured by jealousy, hatred, avarice, or ambition, can never find any rest. The most peaceful spot, the freshest and most calming drinks are alike useless to one who has not freed his heart from the torment of passion.

The soul and the body fall asleep together. As the motion of the blood is calmed, a sweet feeling of peace and quiet spreads through the whole mechanism. The soul feels itself little by little growing heavy as the eyelids droop, and loses its tenseness, as the fibres of the brain relax; thus little by little it becomes as if paralyzed and with it all the muscles of the body. These can no longer sustain the weight of the head, and the soul can no longer bear the burden of thought; it is in sleep as if it were not.

Is the circulation too quick? the soul cannot sleep. Is the soul too much excited? the blood cannot be quieted: it gallops through the veins with an audible murmur/ Such are the two opposite causes of insomnia. A single fright in the midst of our dreams makes the heart beat at double speed and snatches us from needed and delicious repose, as a real grief or an urgent need would do. Lastly as the mere cessation of the functions of the soul produces sleep, there are, even when we are awake (or at least when we are half awake), kinds of very frequent short naps of the mind, vergers' dreams, which show that the soul does not always wait for the body to sleep. For if the soul is not fast asleep, it surely is not far from sleep, since it cannot point out a single object to which it has attended, among the uncounted number of confused ideas which, so to speak, fill the atmosphere of our brains like clouds.

Opium is too closely related to the sleep it produces, to be left out of consideration here. This drug intoxicates, like wine, coffee, etc., each in its own measure and according to the dose. It makes a man happy in a state which would seemingly be the tomb of feeling, as it is the image of death. How sweet is this lethargy! The soul would long never to emerge from it. For the soul has been a prey to the most intense sorrow, but now feels only the joy of suffering past, and of sweetest peace. Opium alters even the will, forcing the soul which wished to wake and to enjoy life, to sleep in spite of itself. I shall omit any reference to the effect of poisons.

Coffee, the well-known antidote for wine, by scourging the imagination, cures our headaches and scatters our cares without laying up for us, as wine does, other headaches for the morrow. But let us contemplate the soul in its other needs.

The human body is a machine which winds its own springs. It is the living image of perpetual movement. Nourishment keeps up the movement which fever excites. Without food, the soul pines away, goes mad, and dies exhausted. The soul is a taper whose light flares up the moment before it goes out. But nourish the body, pour into its veins life-giving juices and strong liquors, and then the soul grows strong like them, as if arming itself with a proud courage, and the soldier whom water would have made to flee, grows bold and runs joyously to death to the sound of drums. Thus a hot drink sets into stormy movement the blood which a cold drink would have calmed.

What power there is in a meal! Joy revives in a sad heart, and infects the souls of comrades, who express their delight in the friendly songs in which the Frenchman excels. The melancholy man alone is dejected, and the studious man is equally out of place [in such company].

Raw meat makes animals fierce, and it would have the same effect on man. This is so true that the English who eat meat red and bloody, and not as well done as ours, seem to share more or less in the savagery due to this kind of food, and to other causes which can be rendered ineffective by education only. This savagery creates in the soul, pride, hatred, scorn of other nations, indocility and other sentiments which degrade the character, just as heavy food makes a dull and heavy mind whose usual traits are laziness and indolence.

Pope understood well the full power of greediness when he said:

Catius is ever moral, ever grave
Thinks who endures a knave is next a knave,
Save just at dinner – then prefers no doubt
A rogue with ven'son to a saint without.
Elsewhere he says:
See the same man in vigor, in the gout,
Alone, in company, in place or out,
Early at business and at hazard late,
Mad at a fox chase, wise at a debate,
Drunk at a borough, civil at a ball,
Friendly at Hackney, faithless at White Hall.

In Switzerland we had a bailiff by the name of M. Steigner de Wittghofen. When he fasted he was a most upright and even a most indulgent judge, but woe to the unfortunate man whom he found on the culprit's bench after he had had a large dinner! He was capable of sending the innocent like the guilty to the gallows.

We think we are, and in fact we are, good men, only as we are gay or brave; everything depends on the way our machine is running. One is sometimes inclined to say that the soul is situated in the stomach, and that Van Helmont, who said that the seat of the soul was in the pylorus, made only the mistake of taking the part for the whole.

To what excesses cruel hunger can bring us! We no longer regard even our own parents and children. We tear them to pieces eagerly and make horrible banquets of them; and in the fury with which we are carried away, the weakest is always the prey of the strongest.

La grossesse, cette émule dsire des pâles couleurs, ne se contente pas d'amener le plus souvent à sa suites le goûts dépravés qui accompagnent ces deux états: elle a quelquefois fait exécuter à l'âme les plus affreux complots; effets d'une maine subite, qui étouffe jusqu'à la loi naturelle. Ce'st ainsi que le cerveau, cette matrice de l'esprit, se pervertit à sa manière, avec celle du corps.

Quelle autre fureur d'homme ou de femme, dans ceux que la continence et la santé poursuivent! C'est peu pour cette fille timide et modeste d'avoir perdu toute honte et toute pudeur; elle ne regarde plus l'inceste, que comme une femme galante regarde l'adultère. Si ses besoins ne trouvent pas de prompts soulagements, ils ne se borneront point aux simples accidents d'une passion utérine, à la manie, etc.; cette malheureuse mourra d'un mal, dont il y a tant de médecins.

One needs only eyes to see the necessary influence of old age on reason. The soul follows the progress of the body, as it does the progress of education. In the weaker sex, the soul accords also with delicacy of temperament, and from this delicacy follow tenderness, affection, quick feelings due more to passion than to reason, prejudices, and superstitions, whose strong impress can hardly be effaced. Màn, on the other hand, whose brain and nerves partake of the firmness of all solids, has not only stronger features but also a more vigorous mind. Education, which women lack, strengthens his mind still more. Thus with such help of nature and art, why should not a man be more grateful, more generous, more constant in friendship, stronger in adversity? But, to follow almost exactly the thought of the author of the Lettres sur la Physiognomie, the sex which unites the charms of the mind and of the body with almost all the tenderest and most delicate feelings of the heart, should not envy us the two capacities which seem to have been given to man, the one merely to enable him better to fathom the allurements of beauty, and the other merely to enable him to minister better to its pleasure.

It is no more necessary to be just as great a physiognomist as this author, in order to guess the quality of the mind from the countenance or the shape of the features, provided these are sufficiently marked, than it is necessary to be a great doctor to recognize a disease accompanied by all it marked symptoms. Look at the portraits of Locke, of Steele, of Boerhaave, of Maupertuis, and the rest, and you will not be surprised to find strong faces and eagle eyes. Look over a multitude of others, and you can always distinguish the man of talent from the man of genius, and often even an honest man from a scoundrel. For example it has been noticed that a celebrated poet combines (in his portrait) the look of a pickpocket with the fire of Prometheus.

History provides us with a noteworthy example of the power of temperature. The famous Duke of Guise was so strongly convinced that Henry the Third, in whose power he had so often been, would never dare assassinate him, that he went to Blois. When the Chancellor Chiverny learned of the duke's departure, he cried, "He is lost." After this fatal prediction had been fulfilled by the event, Chiverny was asked why he made it. "I have known the king for twenty years," said he; "he is naturally kind and even weakly indulgent, but I have noticed that when it is cold, it takes nothing at all to provoke him and send him into a passion."

One nation is of heavy and stupid wit, and another quick, light and penetrating. Whence comes this difference, if not in part from the difference in foods, and difference in inheritance, and in part from the mixture of the diverse elements which float around in the immensity of the void? The mind, like the body, has its contagious diseases and its scurvy.

Such is the influence of climate, that a man who goes from one climate to another, feels the change, in spite of himself. He is a walking plant which has transplanted itself; if the climate is not the same, it will surely either degenerate or improve.

Furthermore, we catch everything from those with whom we come in contact; their gestures, their accent, etc.; just as the eyelid is instinctively lowered when a blow is foreseen, or (as for the same reason) the body of the spectator mechanically imitates, in spite of himself, all the motions of a good mimic.

From what I have just said, it follows that a brilliant man is his own best company, unless he can find others of the same sort. In the society of the unintelligent, the mind grows rusty for lack of exercise, as at tennis a ball that is served badly is badly returned. I should prefer an intelligent man without an education, if he were still young enough, to a man badly educated. A badly trained mind is like an actor whom the provinces have spoiled.

Thus, the diverse states of the soul are always correlative with those of the body. But the better to show this dependence, in its completeness and its causes, let us here make use of comparative anatomy;

let us lay bare the organs of man and of animals. How can human nature be known, if we may not derive any light from an exact comparison of the structure of man and of animals?

In general, the form and the structure of the brains of quadrupeds are almost the same as those of the brain of man; the same shape, the same arrangement everywhere, with this essential difference, that of all the animals man is the one whose brain is largest, and, in proportion to its mass, more convoluted than the brain of any other animal; then come the monkey, the beaver, the elephant, the dog, the fox, the cat. These animals are most like man, for among them, too, one notes the same progressive analogy in relation to the corpus callosum in which Lancisi anticipating the late M. de la Peyronie - established the seat of the soul. The latter, however, illustrated the theory by innumerable experiments. Next after all the quadrupeds, birds have the largest brains. Fish have large heads, but these are void of sense, like the heads of many men. Fish have no corpus callosum, and very little brain, while insects entirely lack brain.

I shall not launch out into any more detail about the varieties of nature, nor into conjectures concerning them, for there is an infinite number of both, as any one can see by reading no further than the treatises of Willis De Cerebro and De Anima Brutorum.

I shall draw the conclusions which follow clearly from these incontestable observations: 1st, that the fiercer animals are, the less brain they have; 2d, that this organ seems to increase in size in proportion to the gentleness of the animal; 3d, that nature seems here eternally to impose a singular condition, that the more one gains in intelligence the more one loses in instinct. Does this bring gain or loss?

Do not think, however, that I wish to infer by that, that the size alone of the brain, is enough to indicate the degree of tameness in animals: the quality must correspond to the quantity, and the solids and liquids must be in that due equilibrium which constitutes health.

If, as is ordinarily observed, the imbecile does not lack brain, his brain will be deficient in its consistency - for instance, in being too soft. The same thing is true of the insane, and the defects of their brains do not always escape our investigation. But if the causes of imbecility, insanity, etc., are not obvious, where shall we look for the causes of the diversity of all minds? They would escape the eyes of a lynx and of an argus. A mere nothing, a tiny fiber, something that could never be found by the most delicate anatomy, would have made of Erasmus and Fontenelle two idiots, and Fontenelle himself speaks of this very fact in one of his best dialogues.

Willis has noticed in addition to the softness of the brain-substance in children, puppies and birds, that the corpora striata are obliterated and discolored in all these animals, and that the striations are as imperfectly formed as in paralytics. Il ajoute, ce qui est vrai, que l'homme a la protubérance annulaire fort grosse; et ensuite toujours diminutivement par dégrés, le singe et les autres animaux nommés ci-devant, tandis que le veau, le boeuf, le loup, la brebis, le cochon, etc. qui ont cette partie d'un tès petit volume, ont les nattes et testes fort gros.

However cautious and reserved one may be about the consequences that can be deduced from these observations, and from many others concerning the kind of variation in the organs, nerves, etc., [one must admit that] so many different varieties cannot be the gratuitous play of nature. They prove at least the necessity for a good and vigorous physical organization, since throughout the animal kingdom the soul gains force with the body and acquires keenness, as the body gains strength.

Let us pause to contemplate the varying capacities of animals to learn. Doubtless the analogy best framed leads the mind to think that the causes we have mentioned produce all the difference that is found between animals and men, although we must confess that our weak understanding, limited to the coarsest observations, cannot see the bonds that exist between cause and effect. This is a kind of harmony that philosophers will never know.

Among animals, some learn to speak and sing; they remember tunes, and strike the notes as exactly as a musician. Others, for instance the ape, show more intelligence, and yet cannot learn music. What is the reason for this, except some defect in the organs of speech? But is this defect so essential to the structure that it could never be remedied? In a word, would it be absolutely impossible to teach the ape a language? I do not think so.

I should choose a large ape in preference to any other, until by some good fortune another kind should be discovered, more like us, for nothing prevents there being such a one in regions unknown to us. The ape resembles us so strongly that naturalists have called it "wild man" or "man of the woods." I should take it in the condition of the pupils of Amman, that is to say, I should not want it to be too young or too old; for apes that are brought to Europe are usually too old. I would choose the one with the most intelligent face, and the one which, in a thousand little ways, best lived up to its look of intelligence. Finally not considering myself worthy to be his master, I should put him in the school of that excellent teacher whom I have just named, or with another teacher equally skillful, if there is one.

You know by Amman's work, and by all those who have interpreted his method, all the wonders he has been able to accomplish for those born deaf. In their eyes he discovered ears, as he himself explained, and in how short a time! In short he taught them to hear, speak, read, and write. I grant that a deaf person's eyes see more clearly and are keener than if he were not deaf, for the loss of one member or sense can increase the strength or acuteness of another, but apes see and hear, they understand what they hear and see, and grasp so perfectly the signs that are made to them, that I doubt not that they would surpass the pupils of Amman in any other game or exercise. Why then should the education of monkeys be impossible? Why might not the monkey, by dint of great pains, at last imitate after the manner of deaf mutes, the motions necessary for pronunciation. I do not dare decide whether the monkey's organs of speech, however trained, would be incapable of articulation. But, because of the great analogy between ape and man and because there is no known animal whose external and internal organs so strikingly resemble man's, it would surprise me if speech were absolutely impossible to the ape. Locke, who was certainly never suspected of credulity, found no difficulty in believing the story told by Sir William Temple in his memoirs, about a parrot which could answer rationally, and which had learned to carry on a kind of connected conversation, as we do. I know that people have ridiculed this great metaphysician; but suppose some one should have announced that reproduction sometimes take place without eggs or a female, would he have found many partisans? Yet M. Trembley has found cases where reproduction takes place without copulation and by fission. Would not Amman too have passed for mad if he had boasted that he could instruct scholars like his in so short a time, before he had happily accomplished the feat? His successes, have, however, astonished the world; and he, like the author of The History of the Polyps, has risen to immortality at one bound. Whoever owes the miracles that he works to his own genius surpasses, in my opinion, the man who owes his to chance. He who has discovered the art of adorning the most beautiful of kingdoms [of nature], and of giving it perfections that it did not have, should be ranked above an idle creator of frivolous systems, or a painstaking author of sterile discoveries. Amman's discoveries are certainly of a much greater value; he has freed men from the instinct to which they seemed to be condemned, and has given them ideas, intelligence, or in a word, a soul which they would never have had. What greater power than this!

Let us not limit the resources of nature; they are infinite, especially when reinforced by great art.

Could not the device which opens the Eustachian canal of the deaf, open that of apes? Might not a happy desire to imitate the master's pronunciation, liberate the organs of speech in animals that imitate so many other signs with such skill and intelligence? Not only do I defy any one to name any really conclusive experiment which proves my view impossible and absurd; but such is the likeness of the structure and functions of the ape to ours that I have very little doubt that if this animal were properly trained he might at last be taught to pronounce, and consequently to know, a language. Then he would no longer be a wild man, nor a defective man, but he would be a perfect man, a little gentleman, with as much matter or muscle as we have, for thinking and profiting by his education.

The transition from animals to man is not violent, as true philosophers will admit. What was man before the invention of words and the knowledge of language? An animal of his own species with much less instinct than the others. In those days, he did not consider himself king over the other animals, nor was he distinguished from the ape, and from the rest, except as the ape itself differs from the other animals, i.e., by a more intelligent face. Reduced to the bare intuitive knowledge of the Leibnizians he saw only shapes and colors, without being able to distinguish between them: the same, old as young, child at all ages, he lisped out his sensations and his needs, as a god that is hungry or tired of sleeping, asks for something to eat, or for a walk.

Words, languages, laws, sciences, and the fine arts have come, and by them finally the rough diamond of our mind has been polished. Man has been trained in the same way as animals. He has become an author, as they have become beasts of burden. A geometrician has learned to perform the most difficult demonstrations and calculations, as a monkey has learned to take his little hat off and on, and to mount his tame dog. All has been accomplished through signs, every species has learned what it could understand, and in this way men have acquired symbolic knowledge, still so called by our German philosophers.

Nothing, as any one can see, is so simple as the mechanism of our education. Everything may be reduced to sounds or words that pass from the mouth of one through the ears of another into his brain. At the same moment, he perceives through his eyes the shape of the bodies of which these words are the arbitrary signs.

But who was the first to speak? Who was the first teacher of the human race? Who invented the means of utilizing the plasticity of our organism? I cannot answer: the names of these first splendid geniuses have been lost in the night of time. But art is the child of nature, so nature must have long preceded it.

We must think that the men who were the most highly organized, those on whom nature has lavished her richest gifts, taught the others. They could not have heard a new sound for instance, nor experienced new sensations, nor been struck by all the varied and beautiful objects that compose the ravishing spectacle of nature without finding themselves in the state of mind of the deaf man of Chartres, whose experience was first related by the great Fontenelle, when, at forty years, he heard for the first time, the astonishing sound of bells.

Would it be absurd to conclude from this that the first mortals tried after the manner of this deaf man, or like animals and like mutes (another kind of animals), to express their new feeling by motions depending on the nature of their imagination, and therefore afterwards by spontaneous sounds, distinctive of each animal, as the natural expression of their surprise, their joy, their ecstasies and their needs? For doubtless those whom nature endowed with finer feeling had also greater facility in expression.

That is the way in which, I think, men have used their feeling and their instinct to gain intelligence and then have employed their intelligence to gain knowledge. Those are the ways, so far as I can understand them, in which men have filled the brain with the ideas, for the reception of which nature made it. Nature and man have helped each other; and the smallest beginnings have, little by little, increased, until everything in the universe could be as easily described as a circle.

As a violin string or a harpsichord key vibrates and gives forth sound, so the cerebral fibers, struck by waves of sound, are stimulated to render or repeat the words that strike them. And as the structure of the brain is such that when eyes well formed for seeing, have once perceived the image of objects, the brain can not help seeing their images and their differences, so when the signs of these differences have been traced or imprinted in the brain, the soul necessarily examines their relations - an examination that would have been impossible without the discovery of signs or the invention of language. At the time when the universe was almost dumb, the soul's attitude toward all objects was that of a man without any idea of proportion toward a picture or a piece of sculpture, in which he could distinguish nothing; or the soul was like a little child (for the soul was then in its infancy) who, holding in his hand small bits of straw or wood, sees them in a vague and superficial way without being able to count or distinguish them. But let some one attach a kind of banner, or standard, to this bit of wood (which perhaps is called a mast), and another banner to another similar object; let the first be known by the symbol 1, and the second by the symbol or number 2, then the child will be able to count the objects, and in this way he will learn all of arithmetic. As soon as one figure seems equal to another in its numerical sing, he will decide without difficulty that they are two different bodies, that 1+1 make 2, and 2+2 make 4, etc.

This real or apparent likeness of figures is the fundamental basis of all truths and of all we know. Among these sciences, evidently those whose signs are less simple and less sensible are harder to understand than the others, because more talent is required to comprehend and combine the immense number of words by which such sciences express the truths in their province. On the other hand, the sciences that are expressed by the numbers or by other small signs, are easily learned; and without doubt this facility rather than its demonstrability is what has made the fortune of algebra.

All this knowledge, with which vanity fills the balloon-like brains of our proud pedants, is therefore but a huge mass of words and figures, which form in the brain all the marks by which we distinguish and recall objects. All our ideas are awakened after the fashion in which the gardener who knows plants recalls all stages of their growth at sight of them. These words and the objects designated by them are so connected in the brain that it is comparatively rare to imagine a thing without the name or sign that is attached to it.

I always use the word "imagine," because I think that everything is the work of imagination, and that all the faculties of the soul can be correctly reduced to pure imagination in which they all consist. Thus judgment, reason, and memory are not absolute parts of the soul, but merely modifications of this kind of medullary screen upon which images of the objects painted in the eye are projected as by a magic lantern.

But if such is the marvelous and incomprehensible result of the structure of the brain, if everything is perceived and explained by imagination, why should we divide the sensitive principle which thinks in man? Is not this a clear inconsistency in the partisans of the simplicity of the mind? For a thing that is divided can no longer without absurdity be regarded as indivisible. See to what one is brought by the abuse of language and by those fine words (spirituality, immateriality, etc.) used haphazard and not understood even by the most brilliant.

Nothing is easier than to prove a system based, as this one is, on the intimate feeling and personal experience of each individual. If the imagination, or let us say, that fantastic part of the brain whose nature is as unknown to us as its way of acting, be naturally small or weak, it will hardly be able to compare the analogy or the resemblance of its ideas, it will be able to see only what is face to face with it, or what affects it very strongly; and how will it see all this! Yet it is always imagination which apperceives, and imagination which represents to itself all objects along with their names and symbols; and thus, once again, imagination is the soul, since it plays all the roles of the soul. By the imagination, by its flattering brush, the cold skeleton of reason takes on living and ruddy flesh, by the imagination the sciences flourish, the arts are adorned, the wood speaks, the echoes sigh, the rocks weep, marble breathes, and all inanimate objects gain life. It is imagination again which adds the piquant charm of voluptuousness to the tenderness of an amorous heart; which makes tenderness bud in the study of the philosopher and of the dusty pedant, which, in a word, creates scholars as well as orators and poets. Foolishly decried by some, vainly praised by others, and misunderstood by all; it follows not only in the train of the graces and of the fine arts, it not only describes but can also measure nature. It reasons, judges, analyzes, compares, and investigates. Could it feel so keenly the beauties of the pictures drawn for it, unless it discovered their relations? No, just as it cannot turn its thoughts on the pleasures of the senses, without enjoying their perfection or their voluptuousness, it cannot reflect on what it has mechanically conceived, without thus being judgment itself.

The more the imagination or the poorest talent is exercised, the more it gains in embonpoint, so to speak, and the larger it grows. It becomes sensitive, robust, broad, and capable of thinking. The best of organisms has need of this exercise.

Man's preeminent advantage is his organism. In vain all writers of books on morals fail to regard as praiseworthy those qualities that come by nature, esteeming only the talents gained by dint of reflection and industry. For whence come, I ask, skill, learning, and virtue, if not from a disposition that makes us fit to become skillful, wise, and virtuous? And whence again, comes this disposition, if not from nature? Only though nature do we have any good qualities; to her we owe all that we are. Why then should I not esteem men with good natural qualities as much as men who shine by acquired and as it were borrowed virtues? Whatever the virtue may be, from whatever source it may come, it is worthy of esteem; the only question is, how to estimate it. Mind, beauty, wealth, nobility, although the children of chance, all have their own value, as skill, learning and virtue all have theirs. Those upon whom nature has heaped her most costly gifts should pity those to whom these gifts have been refused; but, in their character of experts, they may feel their superiority without pride. A beautiful woman would be as foolish to think herself ugly, as an intelligent man to think himself a fool. An exaggerated modesty (a rare fault, to be sure) is a kind of ingratitude towards nature. An honest pride, on the contrary, is the mark of a strong and beautiful soul, revealed by manly features moulded by feeling.

If one's organism is an advantage, and the preeminent advantage, and the source of all others, education is the second. The best made brain would be a total loss without it, just as the best constituted man would be but a common peasant, without knowledge of the ways of the world. But, on the other hand, what would be the use of the most excellent school, without a matrix perfectly open to the entrance and conception of ideas? Il est aussi impossible de donner une seule idée à un homme privé de tous les sens, que de faire un enfant à une femme à laquelle la nature aurait poussé la distraction jusqu'à oublier de faire une vulve, comme je l'ai vu dans une, qui n'avait ni fente, ni vagin, ni matrice, et qui pour cette raison fut démariée après dix ans de mariage.

But if the brain is at the same time well organized and well educated, it is a fertile soil, well sown, that brings forth a hundredfold what it has received: or (to leave the figures of speech often needed to express what one means, and to add grace to truth itself) the imagination, raised by art to the rare and beautiful dignity of genius, apprehends exactly all the relations of the ideas it has conceived, and takes in easily an astounding number of objects, in order to deduce from them a long chain of consequences, which are again but new relations, produced by a comparison with the first, to which the soul finds a perfect resemblance. Such is, I think, the generation of intelligence. I say "finds" as I before gave the epithet "apparent" to the likeness of objects, not because I think that our senses are always deceivers, as Father Malebranche has claimed, or that our eyes, naturally a little unsteady, fail to see objects as they are in themselves (though microscopes prove this to us every day) but in order to avoid any dispute with the Pyrrhonians, among whom Bayle is well known.

I say of truth in general what M. de Fontenelle says of certain truths in particular, that we must sacrifice it in order to remain on good terms with society. And it accords with the gentleness of my character, to a void all disputes unless to what conversation [!]. The Cartesians would here in vain make an onset upon me with their innate ideas. I certainly would not give myself a quarter of the trouble that M. Locke took, to attack such chimeras. In truth, what is the use of writing a ponderous volume to prove a doctrine which became an axiom three thousand years ago?

According to the principles which we have laid down, and which we consider true; he who has the most imagination should be regarded as having the most intelligence or genius, for all these words are synonymous; and again, only by a shameful abuse [of terms] do we think that we are saying different things, when we are merely using different words, different sounds, to which no idea or real distinction is attached.

The finest, greatest or strongest imagination is then the one most suited to the sciences as well as to the arts. I do not pretend to say whether more intellect is necessary to excel in the art of Aristotle or of Descartes than to excel in that of Euripides or of Sophocles, and whether nature has taken more trouble to make Newton than to make Corneille, though I doubt this. But it is certain that imagination alone, differently applied, has produced their diverse triumphs and their immortal glory.

If one is known as having little judgment and much imagination, this means that the imagination has been left too much alone, has, as it were, occupied most of the time in looking at itself in the mirror of its sensations, has not sufficiently formed the habit of examining the sensations themselves attentively. [It means that the imagination] has been more impressed by images than by their truth or the likeness.

Truly, so quick are the responses of the imagination that if attention, that key or mother of the sciences, does not do its part, imagination can do little more than run over and skim its objects.

See that bird on the bough: it seems always ready to fly away. Imagination is like the bird, always carried onward by the turmoil of the blood and the animal spirits. One wave leaves a mark, effaced by the one that follows;

the soul pursues it, often in vain: it must expect to regret the loss of that which it has not quickly enough seized and fixed. Thus, imagination, the true image of time, is being ceaselessly destroyed and renewed.

Such is the chaos and the continuous quick succession of our ideas: they drive each other away even as one wave yields to another. Therefore, if imagination does not, as it were, use one set of its muscles to maintain a kind of equilibrium with the fibers of the brain, to keep its attention for a while upon an object that is on the point of disappearing, and to prevent itself from contemplating prematurely another object – [unless the imagination does all this], it will never be worthy of the fine name of judgment. It will express vividly what it has perceived in the same fashion: it will create orators, musicians, painters, poets, but never a single philosopher. On the contrary, if the imagination be trained from childhood to bridle itself and to keep from being carried away by its own impetuosity – an impetuosity which creates only brilliant enthusiasts – and to check, to restrain, its ideas, to examine them in all their aspects in order to see all sides of an object, then the imagination, ready in judgment, will comprehend the greatest possible sphere of objects, through reasoning; and its vivacity (always so good a sign in children, and only needing to be regulated by study and training) will be only a farseeing insight without which little progress can be made in the sciences.

Such are the simple foundations upon which the edifice of logic has been reared. Nature has built these foundations for the whole human race, but some have used them, while others have abused them.

In spite of all these advantages of man over animals, it is doing him honor to place him in the same class. For, truly, up to a certain age, he is more of an animal than they, since at birth he has less instinct. What animal would die of hunger in the midst of a river of milk? Man alone. Like that child of olden time whom a modern writer refers, following Arnobius, he knows neither the foods suitable for him, nor the water that can drown him, nor the fire that can reduce him to ashes. Light a wax candle for the first time under a child's eyes, and he will mechanically put his fingers in the flame as if to find out what is the new thing that he sees. It is at his own cost that he will learn of the danger, but he will not be caught again. Or, put the child with an animal on a precipice, the child alone falls off; he drowns where the animal would save itself by swimming. At fourteen or fifteen years the child knows hardly anything of the great pleasures in store for him, in the reproduction of his species; when he is a youth, he does not know exactly how to behave in a game which nature teaches animals so quickly. He hides himself as if he were ashamed of taking pleasure, and of having been made to be happy, while animals frankly glory in being Cynics. Without education, they are without prejudices. For one more example, let us observe a dog and a child who have lost their master on a highway: the child cries and does not know to what saint to pray, while the dog, better helped by his sense of smell than the child by his reason, soon finds his master.

Thus nature made us to be lower than animals or at least to exhibit all the more, because of that native inferiority, the wonderful efficacy of education which alone raises us from the level of the animals and lifts us above them. But shall we grant this same distinction to the deaf and to the blind, to imbeciles, madmen, or savages, or to those who have been brought up in the woods with animals; to those who have lost their imagination through melancholia, or in short to all those animals in human form who give evidence of only the rudest instinct? No, all these, men of body but not of mind, do not deserve to be classed by themselves.

We do not intend to hide from ourselves the arguments that can be brought forward against our belief and in favor of a primitive distinction between men and animals. Some say that there is in man a natural law, a knowledge of good and evil, which has never been imprinted on the heart of animals.

But is this objection, or rather this assertion, based on observation? An assertion unfounded on observation may be rejected by a philosopher. Have we ever had a single experience which convinces us that man alone has been enlightened by a ray denied all other animals? If there is no such experience, we can no more know what goes on in animals' minds or even in the minds of other men, than we can help feeling what affects the inner part of our own being. We know that we think, and we feel remorse – an intimate feeling forces us to recognize this only too well; but this feeling in us is insufficient to enable us to judge the remorse of others. That is why we have to take others at their words, or judge them by the sensible and external signs we have noticed in ourselves when we experienced the same accusations of conscience and the same torments.

In order to decide whether animals which do not talk have received the natural law, we must, therefore, have recourse to those signs to which I have just referred, if any such exist. The facts seem to prove it. A dog that bit the master who was teasing it, seemed to repent a minute afterwards; it looked sad, ashamed, afraid to show itself, and seemed to confess its guilt by a crouching and downcast air. History offers us a famous example of a lion which would not devour a man abandoned to its fury, because it recognized him as its benefactor. How much might it be wished that man himself always showed the same gratitude for kindnesses, and the same respect for humanity! Then we should no longer fear either ungrateful wretches, or wars which are the plague of the human race and the real executioners of the natural law.

But a being to which nature has given such a precocious and enlightened instinct, which judges, combines, reasons, and deliberates as far as the sphere of its activity extends and permits, a being which feels attachment because of benefits received, and which leaving a master who treats it badly goes to seek a better one, a being with a structure like ours, which performs the same acts, has the same passions, the same griefs, the same pleasures, more or less intense according to the sway of the imagination and the delicacy of the nervous organization - does not such a being show clearly that it knows its faults and ours, understands good and evil, and in a word, has consciousness of what it does? Would its soul, which feels the same joys, the same mortification and the same discomfiture which we feel, remain utterly unmoved by disgust when it saw a fellow-creature torn to bits, or when it had itself pitilessly dismembered this fellow-creature? If this be granted, it follows that the precious gift now in question would not have been denied to animals: for since they show us sure signs of repentance, as well as of intelligence, what is there absurd in thinking that beings, almost as perfect machines as ourselves, are, like us, made to understand and to feel nature?

Let no one object that animals, for the most part, are savage beasts, incapable of realizing the evil that they do; for do all men discriminate better between vice and virtue? There is ferocity in our species as well as in theirs. Men who are in the barbarous habit of breaking the natural law are not tormented as much by it, as those who transgress for the first time, and who have not been hardened by the force of habit. The same thing is true of animals as of men – both may be more or less ferocious in temperament, and both become more so by living with others like themselves. But a gentle and peaceful animal which lives among other animals of the same disposition and of gentle nurture, will be an enemy of blood and carnage; it will blush internally at having shed blood. There is perhaps this difference, that since among animals everything is sacrificed to their needs, to their pleasures, to the necessities of life, which they enjoy more than we, their remorse apparently should not be as keen as ours, because we are not in the same state of necessity as they. Custom perhaps dulls and perhaps stifles remorse as well as pleasures.

But I will for a moment suppose that I am utterly mistaken in concluding that almost all the world holds a wrong opinion on this subject, while I alone am right. I will grant that animals, even the best of them, do not know the difference between moral good and evil, that they have no recollection of the trouble taken for them, of the kindness done them, no realization of their own virtues. [I will suppose], for instance, that this lion, to which I, like so many others, have referred, does not remember at all that it refused to kill the man, abandoned to its fury, in a combat more inhuman than one could find among lions, tigers and bears, put together. For our compatriots fight, Swiss against Swiss, brother against brother, recognize each other, and yet capture and kill each other without remorse, because a prince pays for the murder. I suppose in shot that the natural law has not been given to animals. What will be the consequences of this supposition? Man is not moulded from a costlier clay; nature has used but one dough, and has merely varied the leaven. Therefore if animals do not repent for having violated this inmost feeling which I am discussing, or rather if they absolutely lack it, man must necessarily be in the same condition. Farewell then to the natural law and all the fine treatises published about it! The whole animal kingdom in general would be deprived of it. But, conversely, if man cannot dispense with the belief that when health permits him to be himself, he always distinguishes the upright, humane, and virtuous, from those who are not human, virtuous, nor honorable: that it is easy to tell vice from virtue, by the unique pleasure and the peculiar repugnance that seems to be their natural effects, it follows that animals, composed of the same matter, lacking perhaps only one degree of fermentation to make it exactly like man's, must share the same prerogatives of animal nature, and that thus there exists no soul or sensitive substance without remorse. The following considerations will reinforce these observations.

It is impossible to destroy the natural law. The impress of it on all animals is so strong, that I have no doubt that the wildest and most savage have some moments of repentance. I believe that that cruel maid of Chalons in Champagne must have sorrowed for her crime, if she really ate her sister. I think that the sam thing is true of all those who commit crimes, even involuntary or temperamental crimes: true of Gaston of Orleans who could not help stealing; of a certain woman who was subject to the same crime when pregnant, and whose children inherited it; of the woman who, in the same condition, ate her husband; of that other women who killed her children, salted their bodies, and ate a piece of them every day, as a little relish; of that daughter of a thief and cannibal who at twelve years followed in his steps, although she had been orphaned when she was a year old, and had been brought up by honest people; to say nothing of many other examples of which the records of our observers are full, all of them proving that there are a thousand hereditary vices and virtues which are transmitted from parents to children as those of the foster mother pass to the children she nurses. Now, I believe and admit that these wretches do not for the most part feel at the time the enormity of their actions. Bulimia, or canine hunger, for example, can stifle all feeling; it is a mania of the stomach that one is compelled to satisfy, but what remorse must be in store for those women, when the come to themselves and grow sober, and remember the crimes they have committed against those they held most dear! What a punishment for an involuntary crime which they could not resist, of which they had no consciousness whatever! However, this is apparently not enough for the judges. For of these women, of whom I tell, one was cruelly beaten and burned, and another was buried alive. I realize that all this is demanded by the interest of society. But doubtless it is much to be wished that excellent physicians might be the only judges. They alone could tell the innocent criminal from the guilty. If reason is the slave of a depraved or mad desire, how can it control the desire?

But if crime carries with it its own more or less cruel punishment, if the most continued and most barbarous habit cannot entirely blot out repentance in the cruelest hearts, if criminals are lacerated by the very memory of their deeds, why should we frighten the imagination of weak minds, by a hell, by specters, and by precipices of fire even less real than those of Pascal? Why must we have recourse to fables, as an honest pope once said himself, to torment even the unhappy wretches who are executed, because we do not think that they are sufficiently punished by their own conscience, their first executioner? I do not mean to say that all criminals are unjustly punished; I only maintain that those whose will is depraved, and whose conscience is extinguished, are punished enough by their remorse when they come to themselves, a remorse, I venture to assert, from which nature should in this case have delivered unhappy souls dragged on by a fatal necessity.

Criminals, scoundrels, ingrates, those in short without natural feelings, unhappy tyrants who are unworthy of life, in vain take a cruel pleasure in their barbarity, for there are calm moments of reflection in which the avenging conscience arises, testifies against them, and condemns them to be almost ceaselessly torn to pieces at their own hands. Whoever torments men is tormented by himself; and the sufferings that he will experience will be the just measure of those that he has inflicted.

On the other hand, there is so much pleasure in doing good, in recognizing and appreciating what one receives, so much satisfaction in practising virtue, in being gentle, humane, kind, charitable, compassionate and generous (for this one word includes all the virtues), that I consider as sufficiently punished any one who is unfortunate enough not to have been born virtuous.

We were not originally made to be learned; we have become so perhaps by a sort of abuse of our organic faculties, and at the expense of the State which nourishes a host of sluggards whom vanity has adorned with the name of philosophers. Nature has created us all solely to be happy - yes, all of us from the crawling worm to the eagle lost in the clouds. For this cause she has given all animals some share of natural law, a share greater or less according to the needs of each animal's organs when in normal condition.

Now how shall we define natural law? It is a feeling that teaches us what we should not do, because we would not wish it to be done to us. Should I dare add to this common idea, that this feeling seems to me but a kind of fear or dread, as salutary to the race as to the individual; for may it not be true that we respect the purse and life of others, only to save our own possessions, our honor, and ourselves; like those Ixions of Christianity who love God and embrace so many fantastic virtues, merely because they are afraid of hell!

You see that natural law is but an intimate feeling that, like all other feelings (thought included) belongs also to imagination. Evidently, therefore, natural law does not presuppose education, revelation, nor legislator, - provided one does not propose to confuse natural law with civil laws, in the ridiculous fashion of the theologians.

The arms of fanaticism may destroy those who support these truths, but they will never destroy the truths themselves.

I do not mean to call in question the existence of a supreme being; on the contrary it seems to me that the greatest degree of probability is in favor of this belief. But since the existence of this being goes no further than that of any other toward proving the need of worship, it is a theoretic truth with very little practical value. Therefore, since we may say, after such long experience, that religion does not imply exact honesty, we are authorized by the same reasons to think that atheism does not exclude it.

Furthermore, who can be sure that the reason for man's existence is not simply the fact that he exists? Perhaps he was thrown by chance on some spot on the earth's surface, nobody knows how nor why, but simply that he must live and die, like the mushrooms which appear from day to day, or like those flowers which border the ditches and cover the walls.

Let us not lose ourselves in the infinite, for we are not made to have the least idea thereof, and are absolutely unable to get back to the origin of things. Besides it does not matter for our peace of mind, whether matter be eternal or have been created, whether there be or be not a God. How foolish to torment ourselves so much about things which we can not know, and which would not make us any happier even were we to gain knowledge about them!

But, some will say, read all such works as those of Fénelon, of Nieuwentyt, of Abadie, of Berham, of Rais, and the rest. Well! what will they teach me or rather what have they taught me? They are only tiresome repetitions of zealous writers, one of whom adds to the other only verbiage, more likely to strengthen than to undermine the foundations of atheism. The number of evidences drawn from the spectacle of nature does not give these evidences any more force. Either the mere structure of a finger, of an ear, of an eye, a single observation of Malpighi proves all, and doubtless much better than Descartes and Malebranche proved it, or all the other evidences prove nothing. Deists, and even Christians, should therefore be content to point out that throughout the animal kingdom the same aims are pursued and accomplished by an infinite number of different mechanisms, all of them however exactly geometrical. For what stronger weapons could there be with which to overthrow atheists? It is true that if my reason does not deceive me, man and the whole universe seem to have been designed for this unity of aim. The sun, air, water, the organism, the shape of bodies, – everything is brought to a focus in the eye as in a mirror that faithfully presents to the imagination all the objects reflected in it, in accordance with the laws required by the infinite variety of bodies which take part in vision. In ears we find everywhere a striking variety, and yet the difference of structure in men, animals, birds, and fishes, does not produce different uses. All ears are so mathematically made, that they tend equally to one and the same end, namely hearing. But would Chance, the deist asks, be a great enough geometrician to vary thus, at pleasure, the works of which she is supposed to be the author, without being hindered by so great a diversity from gaining the same end? Again, the deist will bring forward as a difficulty those parts of the animal that are clearly contained in it for future use, the butterfly in the caterpillar, man in the sperm, a whole polyp in each of its parts, the valvule in the oval orifice, the lungs in the foetus, the teeth in their sockets, the bones in the fluid from which they detach themselves and (in an incomprehensible manner) harden. And since the partisans of this theory, far from neglecting anything that would strengthen proof, never tire of piling up proof upon proof, they are willing to avail themselves of everything, even of the weakness of the mind in certain cases. Look, they say, at men like Spinoza, Vanini, Desbarreau, and Boindin, apostles who honor deism more than they harm it. The duration of their health was the measure of their unbelief, and one rarely fails, they add, to renounce atheism when the passions, with their instrument, the body, have grown weak.

That is certainly the most that can be said in favor of the existence of God: although the last argument is frivolous in that these conversions are short, and the mind almost always regains its former opinions and acts accordingly, as soon as it has regained or rather rediscovered its strength in that of the body. That is, at least, much more than was said by the physician Diderot, in his Pensées Philosophiques, a sublime work that will not convince a single atheist. What reply can, in truth, be made to a man who says, "We do not know nature; causes hidden in her breast might have produced everything. In your turn, observe the polyp of Trembley: does it not contain in itself the causes which bring about regeneration? Why then would it be absurd to think that there are physical causes by reason of which everything has been made, and to which the whole chain of this vast universe is so necessarily bound and held that nothing which happens, could have failed to happen, – causes, of which we are so invincibly ignorant that we have had recourse to a God, who, as some aver, is not so much as a logical entity? Thus to destroy chance is not to prove the existence of a supreme being, since there may be some other thing which is neither chance nor God - I mean, nature. It follows that the study of nature can only make unbelievers; and the way of thinking of all its more successful investigators proves this."

The weight of the universe therefore far from crushing a real atheist does not even shake him. All these evidences of a creator, repeated thousands and thousands of times, evidence that are placed far above the comprehension of men like us, are self-evident (however far one push the argument) only to the anti-Pyrrhonians, or to those who have enough confidence in their reason top believe themselves capable of judging on the basis of certain phenomena, against which, as you see, the atheist can urge others perhaps equally strong and absolutely opposed. For if we listen to the naturalists again, they will tell us that the very causes which, in a chemist's hands, by a chance combination, made the first mirror, in the hands of nature made the pure water, the mirror of the simple shepherdess; that the motion which keeps the world going could have created it, that each body has taken the place assigned to it by its own nature, that the air must have surrounded the earth, and that iron and the other metals are produced by the internal motions of the earth, for one and the same reason; that the sun is as much a natural product as electricity, that it was not made to warm the earth and its inhabitants, whom it sometimes burns, any more than the rain was made to make the seeds grow, which it often spoils; that the mirror and the water were no more made for people to see themselves in, than were all other polished bodies with this same property; that the eye is in truth a kind of glass in which the soul can contemplate the image of objects as they are presented to it by these bodies, but that it is not proved that this organ was really made expressly for this contemplation, nor purposely placed in its socket, and in short it may well be that Lucretius, the physician Lamy, and all Epicureans both ancient and modern were right when they suggested that the eye sees only because it is formed and placed as it is, and that, given once for all, the same rules of motion followed by nature in the generation and development of bodies, this marvelous organ could not have been formed and placed differently.

Such is the pro and the con, and the summary of those fine arguments that will eternally divide the philosophers. I do not take either side.

''`Non nostrum inter vos tantas compenere lites.'

This is what I said to one of my friends, a Frenchman, as frank a Pyrrhonian as I, a man of much merit, and worthy of a better fate. He gave me a very singular answer in regard to the matter. "It is true," he told me, "that the pro and con should not disturb at all the soul of a philosopher, who sees that nothing is proved with clearness enough to force his consent, and that the arguments offered on one side are neutralized by those of the other. However," he continued, "the universe will never be happy, unless it is atheistic." Here are this wretch's reasons. If atheism, said he, were generally accepted, all the forms of religion would then be destroyed and cut off at the roots. No more theological wars, no more soldiers of religion – such terrible soldiers! Nature infected with a sacred poison, would regain its rights and its purity. Deaf to all other voices, tranquil mortals would follow on the spontaneous dictates of their own being, the only commands which can never be despised with impunity and which alone can lead us to happiness through the pleasant paths of virtue.

Such is natural law: whoever rigidly observes it is a good man and deserves the confidence of all the human race. Whoever fails to follow it scrupulously affects, in vain, the specious exterior of another religion; he is a scamp or a hypocrite whom I distrust.

After this, let a vain people think otherwise, let them dare affirm that even probity is at stake in not believing in revelation, in a word that another religion than that of nature is necessary, whatever it may be. Such an assertion is wretched and pitiable; and so is the good opinion which each one gives us of the religion he has embraced! We do not seek here the votes of the crowd. Whoever raises in his heart altars to superstition, is bound to worship idols and not to thrill to virtue.

But since all the faculties of the soul depend to such a degree on the proper organization of the brain and of the whole body, that apparently they are but this organization itself, the soul is clearly an enlightened machine. For finally, even if man alone had received a share of natural law, would he be any less a machine for that? A few more wheels, a few more springs than in the most perfect animals, the brain proportionally nearer the heart and for this very reason receiving more blood - any one of a number of unknown causes might always produce this delicate conscience so easily wounded, this remorse which is no more foreign to matter than to thought, and in a word all the differences that are supposed to exist here. Could the organism then suffice for everything? Once more, yes; since thought visibly develops with our organs, why should not the matter of which they are composed be susceptible of remorse also, when once it has acquired, with time, the faculty of feeling?

The soul is therefore but an empty word, of which no one has any idea, and which an enlightened man should only use to signify the part in us that thinks. Given the least principle of motion, animated bodies will have all that is necessary for moving, feeling, thinking, repenting, or in a word for conducting themselves in the physical realm, and in the moral realm which depends upon it.

Yet we take nothing for granted; those who perhaps think that all the difficulties have not yet been removed shall now read of experiments that will completely satisfy them.

The flesh of all animals palpitates after death. This palpitation continues longer, the more cold blooded the animal is and the less it perspires. Tortoises, lizards, serpents, etc. are evidence of this.

Muscles separated from the body contract when they are stimulated.

The intestines keep up their peristaltic or vermicular motion for a long time.

According to Cowper, a simple injection of hot water reanimates the heart and the muscles.

A frog's heart moves for an hour or more after it has been removed from the body, especially when exposed to the sun or better still when placed on a hot table or chair. If this movement seem totally lost, one has only to stimulate the heart, and that hollow muscle beats again. Harvey made this same observation on toads.

Bacon of Verulam in his treatise Sylva Sylvarum cites the case of a man convicted of treason, who was opened alive, and whose heart thrown into hot water leaped several times, each time less high, to the perpendicular height of two feet.

Take a tiny chicken still in the egg, cut out the heart and you will observe the same phenomena as before, under almost the same conditions. The warmth of the breath alone reanimates an animal about to perish in the air pump.

The same experiments, which we owe to Boyle and to Stenon, are made on pigeons, dogs, and rabbits. Pieces of their hearts beat as their whole hearts would. The same movements can be seen in paws that have been cut off from moles.

The caterpillar, the worm, the spider, the fly, the eel - all exhibit the same phenomena; and in hot water, because of the fire it contains, the movement of the detached parts increases.

A drunken soldier cut off with one stroke of his sabre an Indian rooster's head. The animal remained standing, then walked, and ran: happening to run against a wall, it turned around, beats its wings still running, and finally fell down. As it lay on the ground, all the muscles of this rooster kept on moving. That is what I saw myself, and almost the same phenomena can easily be observed in kittens or puppies with their heads cut off.

Polyps do more than move after they have been cut in pieces. In a week they regenerate to form as many animals as there are pieces. I am sorry that these facts speak against the naturalists' system of generation; or rather I am very glad of it, for let this discovery teach us never to reach a general conclusion even on the ground of all known (and most decisive) experiments.

Here we have many more facts than are needed to prove, in an incontestable way, that each tiny fiber or part of an organized body moves by a principle which belongs to it. Its activity, unlike voluntary motions, does not depend in any way on the nerves, since the movements in question occur in parts of the body which have no connection with the circulation. But if this force is manifested even in sections of fibers the heart, which is a composite of peculiarly connected fibers, must possess the same property. I did not need Bacon's story to persuade me of this. It was easy for me to come to this conclusion, both from the perfect analogy of the structure of the human heart with that of animals, and also from the very bulk of the human heart, in which this movement escapes our eyes only because it is smothered, and finally because in corpses all the organs are cold and lifeless. If executed criminals were dissected while their bodies are still warm, we should probably see in their hearts the same movements that are observed in the face-muscles of those that have been beheaded.

The motive principle of the whole body, and even of its parts cut in pieces, is such that it produces not irregular movements, as some have thought, but very regular ones, in warm blooded and perfect animals as well as in cold and imperfect ones. No resource therefore remains open to our adversaries but to deny thousands and thousands of facts which every man can easily verify.

If now any one ask me where is this innate force in our bodies, I answer that it very clearly resides in what the ancients called the parenchyma, that is to say, in the very substance of the organs not including the veins, the arteries, the nerves, in a word, that it resides in the organization of the whole body, and that consequently each organ contains within itself forces more or less active according to the need of them.

Let us now go into some detail concerning these springs of the human machine. All the vital, animal, natural, and automatic motions are carried on by their action. Is it not in a purely mechanical way that the body shrinks back when it is struck with terror at the sight of an unforeseen precipice, that the eyelids are lowered at the menace of a blow, as some have remarked, and that the pupil contracts in broad daylight to save the retina, and dilates to see objects in darkness? Is it not by mechanical means that the pores of the skin close in winter so that the cold cannot penetrate to the interior of the blood vessels, and that the stomach vomits when it is irritated by poison, by a certain quantity of opium and by all emetics, etc.? that the heart, the arteries and the muscles contract in sleep as well as in waking hours, that the lungs serve as bellows continually in exercise, n'est-ce pas machinalement qu'agissent tous les sphincters de la vessie, du rectum, etc.? that the heart contracts more strongly than any other muscle? que les muscles érecteurs font dresser la verge dans l'homme, comme dans les animaux qui s'en battent le ventre, et même dans l'enfant, capable d'érection, pour peu que cette partie soit irritée? Ce qui prouve, pour le dire en passant, qu'il est un ressort singulier dans ce membre, encore peu connu, et qui produit des effets qu'on n'a point encoure bien expliqués, malgré toutes les lumières de l'anatomie.

I shall not go into any more detail concerning all these little subordinate forces, well known to all. But there is another more subtle and marvelous force, which animates them all; it is the source of all our feelings, of all our pleasures, of all our passions, and of all our thoughts: for the brain has its muscles for thinking, as the legs have muscles for walking. I wish to speak of this impetuous principle that Hippocrates calls enormon (soul). This principle exists and has its seat in the brain at the origin of the nerves, by which it exercises its control over all the rest of the body. By this fact is explained all that can be explained, even to the surprising effect of maladies of the imagination.

Mais, pour ne pas languir dans une richesse et un fécondité mal entendue, il faut se borner à un petit nombre de questions et de réflexions.

Pourquoi la vue ou la simple idée d'une belle femme nous cause-t-elle des mouvements et des désirs singuliers? Ce qui se passe alors dans certains organes, vient-il de la nature même de ces organes? Point du toutl mais du commerce et de l'espèce de sympathie de ces muscles avec l'imagination. Il n'y a ici qu'un premier ressort excité par le bene placitum des anciens, ou par l'image de la beauté, qui en excite un autre, lequel était fort assoupi, quand l'imagination l'a éveillé: et comment cela, si ce n'est par le désordre et le tumulte du sang et des esprits, qui galopent avec une promptitude extraordinaire, et vont gonfler les corps caverneux?

Puisqu'il est des commincations évidents entre la mère et l'enfant, et qu'il est dur de nier des fair rapportés par Tulpius et par d'autres écrivains aussi dignes de foi (il n'y en a point qui le soient plus), nous croirons que c'est par la même voie que le foetus ressent l'impétuoisité de l'imagination maternelle, comme une cire molle reçe;oit toutes sortes d'impressions; et que les mêmes traces, ou envies de la mère, peuvent s'imprimer sur le foetus, sans que cela puisse se comprendre, quoiqu'en disent Blondel et tous ses adhérenets. Ainsi nous faisons réparation d'honneur au P. Malebranche, beaucoup trop raillé de sa crédulité par les auteurs qui n'ont point observé d'assex près la nature et ont voulu l'assujettir à leur idées.

Look at the portrait of the famous Pope who is, to say the least, the Voltaire of the English. The effort, the energy of his genius are imprinted upon his countenance. It is convulsed. His eyes protrude from their sockets, the eyebrows are raised with the muscles of the forehead. Why? Because the brain is in travail and all the body must share in such a laborious deliverance. If there were not an internal cord which pulled the external ones, whence would come all these phenomena? To admit a soul as explanation of them, is to be reduced to [explaining phenomena by] the operations of the Holy Spirit.

In fact, if what thinks in my brain is not a part of this organ and therefore of the whole body, why does my blood boil, and the fever of my mind pass into my veins, when lying quietly in bed, I am forming the plan of some work or carrying on an abstract calculation? Put this question to men of imagination, to great poets, to men who are enraptured by the felicitous expression of sentiment, and transported by an exquisite fancy or by the charms of nature, of truth, or of virtue! By their enthusiasm, by what they will tell you they have experienced, you will judge the cause by its effects; by that harmony which Borelli, a mere anatomist, understood better than all the Leibnizians, you will comprehend the material unity of man. In short, if the nerve-tension which causes pain occasions also the fever by which the distracted mind looses its willpower, and if, conversely, the mind too much excited, disturbs the body (and kindles that inner fire which killed Bayle while he was still so young)l if an agitation rouses my desire and my ardent wish for what, a moment ago, I cared nothing about, and if in their turn certain brain impressions excite the same longing and the same desires, then why should we regard as double what is manifestly one being? In vain you fall back on the power of the will, since for one order that the will gives, it bows a hundred times to the yoke, And what wonder that in health the body obeys, since a torrent of blood and of animal spirits forces its obedience, and since the will has as ministers an invisible legion of fluids swifter than lightning and ever ready to do its bidding! But as the power of the will is exercised by means of the nerves, it is likewise limited by them. La meilleure volonté d'un amant épuisé, les plus violent desires lui rendront-ils sa vigueur perdue? Hélas! non; et elle en sera la première punie, parce-que, posées certaines circonstances, il n'est pas dans sa puissance de ne pas vouloir du plaisir. Ce que j'ai dit de la paralysie, etc. revient ici.

Does the result of jaundice surprise you? Do you not know that the color of bodies depends on the color of the glasses through which we look at them, and that whatever is the color of the humors, such is the color of objects, at least for us, vain playthings of a thousand illusions? But remove this color from the aqueous humor of the eye, let the bile flow through its natural filter, then the soul having new eyes, will no longer see yellow. Again,. is it not thus, by removing cataract, or by injecting the Eustachian canal, that sight is restored to the blind, or hearing to the deaf? How many people, who were perhaps only clever charlatans, passed for miracle workers in the dark ages! Beautiful the soul, and powerful the will which can not act save by permission of the bodily conditions, and whose tastes change with age and fever! Should we, then, be astonished that philosophers have always had in mind the health of the body, to preserve the health of the soul, that Pythagoras gave rules for the diet as carefully as Plato forbade wine? The regime suited to the body is always the one with which sane physicians think they must begin, when it is a question of forming the mind, and of instructing it in the knowledge of truth and virtue; but these are vain words in the disorder of illness, and in the tumult of the senses. Without the precepts of hygiene, Epictetus, Socrates, Plato, and the rest preach in vain: all ethics is fruitless for one who lacks his share of temperance; it is the source of all virtues, as intemperance is the source of all vices.

Is more needed, (for why lose myself in discussion of the passions which are all explained by the term, enormon, of Hippocrates) to prove that man is but an animal, or a collection of springs which wind each other up, without or being able to tell at what point in this human circle, nature has begun? If these springs differ among themselves, these differences consist only in their position and in their degrees of strength, and never in their nature; wherefore the soul is but a principle of motion or a material and sensible part of the brain, which can be regarded, without fear of error, as the mainspring of the whole machine, having a visible influence on all the parts. The soul seems even to have been made for the brain, so that all other parts of the system are but a kind of emanation from the brain. This will appear from certain observations, made on different embryos, which I shall now enumerate.

This oscillation, which is natural or suited to our machine, and with which each fibre and even each fibrous element, so to speak, seems to be endowed, like that of a pendulum, cannot keep up forever. It must be renewed, as it loses strength, invigorated when it is tired, and weakened when it is disturbed by an excess of strength and vigor. In this alone, true medicine consists.

The body is but a watch, whose watchmaker is the new chyle. Nature's first care, when the chyle enters the blood, is to excite in it a kind of fever which the chemists, who dream only of retorts, must have taken for fermentation. This fever produces a greater filtration of spirits, which mechanically animate the muscles and the heart, as if they had been sent there by order of the will.

These then are the causes or forces of life which thus sustain for a hundred years that perpetual movement of the solids and liquids which is as necessary to the first as to the second. But who can say whether the solids contribute more than the fluids to this movement or vice versa? All that we know is that the action of the former would soon cease without the help of the latter, that is, without the help of the fluids which by their onset rouse and maintain the elasticity of the blood vessels on which their own circulation depends. From this it follows that after death the natural resilience of each substance is still more or less strong according to the remnants of life which it outlives, being the last to perish. So true is it that this force of the animal parts can be preserved and strengthened by that of the circulation, but that it does not depend on the strength of the circulation, since, as we have seen, it can dispense with even the integrity of each member or organ.

I am aware that this opinion has not been relished by all scholars, and that Stahl especially had much scorn for it. This great chemist had wished to persuade us that the soul is the sole cause of all our movements. But this is to speak as a fanatic and not as a philosopher.

To destroy the hypothesis of Stahl, we need not make as great an effort as I find that others have done before me. We need only glance at a violinist. What flexibility, what lightness in his fingers! The movements are so quick, that it seems almost as if there were no succession. But I pray, or rather I challenge, the followers of Stahl who understand so perfectly all that our soul can do, to tell me how it could possibly execute so many motions so quickly, motions, moreover, which take place so far from the soul, and in so many different places. That is to suppose that a flute player could play brilliant cadences on an infinite number of holes that he could not know, and on which he could not even put his finger!

But let us say with M. Hecquet that all men may not go to Corinth. Why should not Stahl have been even more favored by nature as a man than as a chemist and a practioner? Happy mortal, he must have received a soul different from the rest of mankind, - a sovereign soul, which, not content with having some control over the voluntary muscles, easily held the reins of all the movements of the body, and could suspend them, calm them, or excite them at its pleasure! With so despotic a mistress, in whose hands were, in a sense, the beating of the heart, and the laws of circulation, there could certainly be no fever, no pain, no weariness, ni honteuse impuissance, ni facheux priapisme! The soul wills, and the springs play, contract or relax. But how did the springs of Stahl's machine get out of order so soon? He who has in himself so great a doctor, should be immortal.

Moreover, Stahl is not the only one who has rejected the principle of the vibration of organic bodies. Greater minds have not used the principle when they wished to explain the actions of the heart, l'érection du penis, etc. One need only read the Institutions of Medicine by Boerhaave to see what laborious and enticing systems this great man was obliged to invent, by the labor of his mighty genius, through failure to admit that there is so wonderful a force in all bodies.

Willis and Perrault, minds of a more feeble stamp, but careful observers of nature (whereas nature was known to the famous Leyden professor only through others and second hand, so to speak) seem to have preferred to suppose a soul generally extended over the whole body, instead of the principle which we are describing. But according to this hypothesis (which was the hypothesis of Vergil and of all Epicureans, an hypothesis which the history of the polyp might seem at first sight to favor) the movements which go on after the death of the subject in which they inhere are due to a remnant of soul still maintained by the parts that contract, though, from the moment of death, these are not excited by the blood and spirits. Whence it may be seen that these writers, whose solid works easily eclipse all philosophic fables, are deceived only in the manner of those who have endowed matter with the faculty of thinking. I mean to say, by having expressed themselves badly in obscure and meaningless terms. In truth, what is this remnant of a soul, if it is not the "moving force" of the Leibnizians (badly rendered by such an expression), which however Perrault in particular has really foreseen. See his Treatise on the Mechanism of Animals.

Now that it is clearly proved against the Cartesians, the followers of Stahl, the Malebranchists, and the theologians who little deserve to be mentioned here, that matter is self-moved, not only when organized, as in a whole heart, for example, but even when this organization has been destroyed, human curiosity would like to discover how a body, by the fact that it is originally endowed with the breath of life, finds itself adorned in consequence with the faculty of feeling, and thus with that of thought. And, heavens, what efforts have not been made by certain philosophers to manage to prove this! and what nonsense of this subject I have had the patience to read!

All that experience teaches us is that while movement persists, however slight it may be, in one or more fibres, we need only stimulate them to re-excite and animate this movement almost extinguished. This has been shown in the host of experiments with which I have undertaken to crush the systems. It is therefore certain that motion and feeling excite each other in turn, both in a whole body and in the same body when its structure is destroyed, to say nothing of certain plants which seem to exhibit the same phenomena of the union of feeling and motion.

But furthermore, how many excellent philosophers have shown that thought is but a faculty of feeling, and that the reasonable soul is but the feeling soul engaged in contemplating its ideas and in reasoning! This would be proved by the fact alone that when feeling is stifled, thought also is checked, for instance in apoplexy, in lethargy, in catalepsis, etc. For it is ridiculous to suggest that, during these stupors, the soul keeps on thinking, even though it does not remember the ideas that it has had.

As to the development of feeling and motion, it is absurd to waste time seeking for its mechanism. The nature of motion is as unknown to us as that of matter. How can we discover how it is produced unless, like the author of The History of the Soul, we resuscitate the old and unintelligible doctrine of substantial forms? I am then quite as content not to know how inert and simple matter becomes active and highly organized, as not to be able to look at the sun without red glasses; and I am as little disquieted concerning the other incomprehensible wonders of nature, the production of feeling and of thought in a being which earlier appeared to our limited eyes as a mere clod of clay.

Grant only that organized matter is endowed with a principle of motion, which alone differentiates it from the inorganic (and can one deny this in the face of the most incontestable observation?) and that among animals, as I have sufficiently proved, everything depends upon the diversity of this organization: these admissions suffice for guessing the riddle of substances and of man. It thus appears that there is but one type of organization in the universe, and that man is the most perfect example. He is to the ape, and to the most intelligent animals, as the planetary pendulum of Huyghens is to a watch of Julien Leroy. More instruments, more wheels and more springs were necessary to mark the movements of the planets than to mark or strike the hours; and Vaucanson, who needed more skill for making his flute player than for making his duck, would have needed still more to make a talking man, a mechanism no longer to be regarded as impossible, especially in the hands of another Prometheus. In like fashion, it was necessary that nature should use more elaborate art in making and sustaining a machine which for a whole century could mark all motions of the heart and of the mind; for though one does not tell time by the pulse, it is at least the barometer of the warmth and the vivacity by which one may estimate the nature of the soul. I am right! The human body is a watch, a large watch constructed with such skill and ingenuity, that if the wheel which marks the second happens to stop, the minute wheel turns and keeps on going its round, and in the same way the quarter-hour wheel, and all the others go on running when the first wheels have stopped because rusty or, for any reason, out of order. Is it not for a similar reason that the stoppage of a few blood vessels is not enough to destroy or suspend the strength of the movement which is in the heart as in the mainspring of the machine; since, on the contrary, the fluids whose volume is diminished, having a shorter road to travel, cover the ground more quickly, borne on as by a fresh current which the energy of the heart increases in proportion to the resistance it encounters at the ends of the blood-vessels? And is not this the reason why the loss of sight (caused by the compression of the optic nerve and its ceasing to convey the images of objects) no more hinders hearing, than the loss of hearing (caused by the obstruction of the functions of the auditory nerve) implies the loss of sight? In the same way, finally, does not one man hear (except immediately after his attack) without being able to say what he hears, while another who hears nothing, but whose lingual nerves are uninjured in the brain, mechanically tells of all the dreams which pass through his mind? These phenomena do not surprise enlightened physicians at all. They know what to think about man's nature (and more accurately to express myself in passing) of two physicians, the better one and the one who deserves more confidence is always, in my opinion, the one who is more versed in the physique or mechanism of the human body, and who, leaving aside the soul and all the anxieties which this chimera gives to fools and to ignorant men, is seriously occupied only in pure naturalism.

Therefore let the pretended M. Charp deride philosophers who have regarded animals as machines. How different is my view! I believe that Descartes would be a man in every way worthy of respect, if, born in a century that he had not been obliged to enlighten, he had known the value of experiment and observation, and the danger of cutting loose from them. But it is none the less just for me to make an authentic reparation to this great man for all the insignificant philosophers - poor jesters, and poor imitators of Locke - who instead of laughing impudently at Descartes, might better realize that without him the field of philosophy, like the field of science without Newton, might perhaps be still uncultivated.

This celebrated philosopher, it is true, was much deceived, and no one denies that. But at any rate he understood animal nature, he was the first to prove completely that animals are pure machines. And after a discovery of this importance demanding so much sagacity, how can we without ingratitude fail to pardon all his errors!

In my eyes, they are all atoned for by that great confession. For after all, although he extols the distinctness of the two substances, this is plainly but a trick of skill, a ruse of style, to make theologians swallow a poison, hidden in the shade of an analogy which strikes everybody else and which they alone fail to notice. For it is this, this strong analogy, which forces all scholars and wise judges to confess that these proud and vain beings, more distinguished by their pride than by the name of men however much they may wish to exalt themselves, are at bottom only animals and machines which, though upright, go on all fours. They all have this marvelous instinct, which is developed by education into mind, and which always has its seat in the brain (or for want of that when it is lacking or hardened, in the medulla oblongata) and never in the cerebellum; for I have often seen the cerebellum injured, and other observers have found it hardened, when the soul has not ceased to fulfil its functions.

To be a machine, to feel, to think, to know how to distinguish good from bad, as well as blue from yellow, in a word, to be born with an intelligence and a sure moral instinct, and to be but an animal, are therefore characters which are no more contradictory, than to be an ape or a parrot and to be able to give oneself pleasure. Car, puisque l'occasion se présente de le dire, qui eut jamais deviné à priori qu'une goutte de la liqeur qui se lance dans l'accouplement fit ressentir des plaisirs divins, et qu'il en naîtrait une petite créature, qui pourrait un jour, posées certaines lois, jouir des même délices? I believe that thought is so little incompatible with organized matter, that it seems to be one of its properties on a par with electricity, the faculty of motion, impenetrability, extension, etc.

Do you ask for further observations? Here are some which are incontestable and which all prove that man resembles animals perfectly, in his origin as well as in all the points in which we have thought it essential to make the comparison.

J'en appale à la bonne foi de nos observateurs. Qu'ils nous disent s'il ne'st pas vrai que l'homme dans son principe n'est qu'un ver, qui devient homme, comme la chenille paillon. Les plus graves auteurs [Boerhaave, Inst. Med. et tant d'autres] nous ont appris comment il faut s'y prendre pour voir cet animalcule. Tous les curieux l'ont vu, comme Hartsoeker, dans la semence de l'homme, et non dans celle de la femme; il n'y a que le plus adroit, ou le plus vigoreux qui ait la force de s'insinuer et de s'implanter dans l'oeuf que fournit la femme, et qui lui donne sa première nourriture. Cet oeuf, quelquefois surpris dans les trompes de Fallope, est porté par ces canaux à la matrice, où il prend racine, comme un grain de blé dans la terre. Mais quoiqu'il y devienne monstru-eux par sa croissance de 9 mois, il ne diffère point des oeufs des autres femelles, si ce n'est que sa peau (l'amnios) ne se durcit jamais, et se dilate prodigeusement, comme on en peut juger en comparant les foetus trovés en situation et près d'éclore (ce que j'ai eu le plaisir d'observer dans une femme morte un moment avant l'accouchement), avec d'autres petits embryons très proches de leur origine: car alors c'est toujours l'oeuf dans sa coque, et l'animal dans l'oeuf, qui, gêné dans ses mouvements, cherche machinalement à voir le jour; et pour y réussir, il commence par rompre avec la tête cette membrance, d'oû il sort, comme le pulet, l'oiseau, etc., de la leur. J'ajouterai une observation que je ne trouve nulle part; c'est que l'amnios n'en est pas plus mince, pour s'être prodigieusement étendu; semblable en cela à la matrice dont la substance même se gonfle de sucs infiltrés, indépendamment de la réplétion et du déploiement de tous ses coudes vasculeux.

Let us observe man both in and out of his shell, let us examine young embryos of four, six, eight or fifteen days with a microscope; after that time our eyes are sufficient. What do we see? The head alone; a little round egg with two black points which mark the eyes. Before that, everything is formless, and one sees only a medullary pulp, which is the brain, in which are formed first the roots of the nerves, that is, the principle of feeling, and the heart, which already within this substance has the power of beating of itself; it is the punctum saliens of Malpighi, which perhaps already owes a part of its excitability to the influence of the nerves. Then little by little, one sees the head lengthen from the neck, which, in dilating, forms first the thorax inside which the heart has already sunk, there to become stationary; below that is the abdomen which is divided by a partition (the diaphragm). One of these enlargements of the body forms the arms, the hands, the fingers, the nails, and the hair; the other forms the thighs, the legs, the feet, etc., which differ only in their observed situation, and which constitute the support and the balancing pole of the body. The whole process is a strange sort of growth, like that of plants. On the tops of our heads is hair in place of which the plants have leaves and flowers; everywhere is shown the same luxury of nature, and finally the directing principle of plants is placed where we have our soul, that other quintessence of man.

Such is the uniformity of nature, which we are beginning to realize; and the analogy of the animal with the vegetable kingdom, of man with the plant. Perhaps there even are animal plants, which in vegetating, either fight as polyps do, or perform other functions characteristic of animals.

Voilà à peu près tout ce qu'on sait de la génération. Que les parties qui s'attirent, qui sont faites pur s'unir ensemble et pour occuper telle ou telle place, se réunissent toutes suivant leur nature; et qu'ainsi se forment les yeux, le coeur, l'estomac et enfin tout le corps, comme de grans hommes l'ont écrit, cela est possible. Mais, comme l'expérience nous abandonne au milieu des ces subtilités, je ne supposerai rien, regardant tout ce qui ne frappe pas mes sens comme un mystère impénetrable. Il est si rare que les deux emences se rencontrent dans le congrès, que je serais tenté de croire que la semence de la femme est inutile à la génération.

Mais comment en expliquer les phénomènes, sans ce commode rapport de parties, qui rend si bien raison des ressemblances des enfants, tantôt au père, et tantôt à la mère? D'un autre côté, l'embarras d'une explication doit-elle contrebalancer un fait? Il me parait que c'est le mâle qui fait tout, dans une femme qui dorrt, comme dans la plus lubrique. L'arrangement des parties serait done fait de toute éternité dans le germe, ou dans le ver même de l'homme. Mais tout ceci est fourt au-dessus de la portée des plus excellents observateurs. Comme ils n'y peuvent rien saisir, ils ne peuvent pas plus juger de la mécanique de la formation et du développment des corps, qu'une taupe du chemin qu'un cerf peut parcourir.

We are veritable moles in the field of nature; we achieve little more than the mole's journey and it si our pride which prescribes limits to the limitless. We are in the position of a watch that should say (a writer of fables would make the watch a hero in a silly tale): "I was never made by that fool of a workman, I who divide time, who mark so exactly the course of the sun, who repeat aloud the hours which I mark! No! that is impossible!" In the same way, we disdain, ungrateful wretches that we are, this common mother of all kingdoms, as the chemists say. We imagine, or rather we infer, a cause superior to that which we owe all, and which truly has wrought all things in an inconceivable fashion. No; matter contains nothing base, except to the vulgar eyes which do not recognize her in her most splendid works; and nature is no stupid workman. She creates millions of men, with a facility and a pleasure more intense than the effort of a watchmaker in making the most complicated watch. Her power shines forth equally in creating the lowliest insect and in creating the most highly developed man; the animal kingdom costs her no more than the vegetable, and the most splendid genius no more than a blade of wheat. Let us then judge by what we see of that which is hidden from the curiosity of our eyes and of our investigations, and let us not imagine anything beyond. Let us observe the ape, the beaver, the elephant, etc., in their operations. If it is clear that these activities cannot be performed without intelligence, why refuse intelligence to these animals? And if you grant them a soul our are lost, you fanatics! You will in vain say that you assert nothing about the nature of the animal soul and that you deny its immortality. Who does not see that this is a gratuitous assertion; who does not see that the soul of an animal must be either mortal or immortal, whichever ours is, and that it must therefore undergo the same fate as ours, whatever that may be, and that thus in admitting that animals have souls, you fall into Scylla in an effort to avoid Charybdis?

Break the chain of your prejudices, arm yourselves with the torch of experience, and you will render nature the honor she deserves, instead of inferring anything to her disadvantage, from the ignorance in which she has left you. Only open wide your eyes, only disregard what you cannot understand, and you will see that the ploughman whose intelligence and ideas extend no further than the bounds of his furrow, does not differ essentially from the greatest genius, - a truth which the dissection of Descartes's and of Newton's brains would have proved; you will be persuaded that the imbecile and the fool are animals with human faces, as the intelligent ape is a little man in another shape; in short, you will learn that since everything depends absolutely on difference of organization , a well constructed animal which has studied astronomy, can predict an eclipse, as it can predict recovery or death when it has used its genius and its clearness of vision, for a time, in the school of Hippocrates and at the bedside of the sick. By this line of observations and truths, we come to connect the admirable power of thought with matter, without being able to see the links, because the subject of this attribute is essentially unknown to us.

Let us not say that every machine or every animal perishes altogether or assumes another form after death, for we know absolutely nothing about the subject. On the other hand, to assert that an immortal machine is a chimera or a logical fiction, is to reason as absurdly as caterpillars would reason if, seeing the castoff skins of their fellow caterpillars, they should bitterly deplore the fate of their species, which to them would seem to come to nothing. The soul of these insects (for each animal has its own) is too limited to comprehend the metamorphoses of nature. Never one of the most skillful among them could have imagined that it was destined to become a butterfly. It is the same way with us. What more do we know of our destiny than of our origin? Let us then submit to an invincible ignorance on which our happiness depends.

He who so thinks will be wise, just, tranquil about his fate, and therefore happy. He will await death without either fear or desire, and will cherish life (hardly understanding how disgust can corrupt a heart in this place of many delights); he will be filled with reverence, gratitude, affection, and tenderness for nature, in proportion to his feeling of the benefits he has received from nature; he will be happy, in short, in feeling nature, and in being present at the enchanting spectacle of the universe, and we will surely never destroy nature either in himself or in others. More than that! Full of humanity, this man will love human character even in his enemies. Judge how he will treat others. He will pity the wicked without hating them; in his eyes, they will be but mis-made men. But in pardoning the faults of the structure of mind and body, he will none the less admire the beauties and the virtues of both. Those whom nature shall have favored will seem to him to deserve more respect than those whom she has treated in step-motherly fashion. Thus, as we have seen, natural gifts, the source of all acquirements, gain from the lips and heart of the materialist, the homage which every other thinker unjustly refuses them. In short, the materialist, convinced, in spite of the protests of his vanity, that is he but a machine or an animal, will not maltreat his kind, for he will know too well the nature of those actions, whose humanity is always in proportion to the degree of analogy proved above [between human beings and animals]; and following the natural law given to all animals, he will not wish to do to others what he would not wish them to do to him.

Let us then conclude boldly that man is a machine, and that in the whole universe there is but a single substance differently modified. This is no hypothesis set forth by dint of a number of postulates and assumptions; it is not the work of prejudice, nor even of my reason alone; I should have disdained a guide which I think to be so untrustworthy, had not my senses, bearing a torch, so to speak, induced me to follow reason by lighting the way themselves. Experience has thus spoken to me in behalf of reason; and in this way I have combined the two.

But it must have been noticed that I have not allowed myself even the most vigorous and immediately deduced reasoning, except as a result of a multitude of observations which no scholar will contest; and furthermore, I recognize only scholars as judges of the conclusions which I draw from the observations; and I hereby challenge every prejudiced man who is neither anatomist, nor acquainted with the only philosophy which can here be considered, that of the human body. Against so strong and solid an oak, what could the weak reeds of theology, of metaphysics, and of the schools, avail, – childish arms, like our parlor foils, that may well afford the pleasure of fencing, but can never wound an adversary. Need I say that I refer to the empty and trivial notions, to the pitiable and trite arguments that will be urged (as long as the shadow of prejudice or of superstition remains on earth for the suppose incompatibility of two substances which meet and move each other unceasingly? Such is my system, or rather the truth, unless I am much deceived. It is short and simple. Dispute it now who will.

www.ReadHowYouWant.com

You can buy our **Large Type** and **EasyRead** books from our www.ReadHowYouWant.com website, from websites like Amazon.com and through your UK and North American bookshop.

EasyRead books are designed to make your reading easy and enjoyable. **EasyRead** books are published in different font sizes, so you can select the font size best for you.

EasyRead is for people with normal eyesight who want books in an easy-to-read format.

EasyRead Comfort is for people who find reading small print tiring but do not need large print.

EasyRead Large is for people who find it easier to read larger print.

The **EasyRead** font, character, word and line spacing have all been set to make it as easy as possible both to recognize a word, and to run your eye along a line of text without losing your place. We split as few words as possible at line ends, as split words make reading harder and can be annoying. For out-of-copyright books, words have been changed to modern spelling (where the sense is not affected), and the original hyphenation of compound words has been retained where this improves word recognition or sense.

With most things you buy, you get a choice, and you can choose the make and model that suits best. There is a big exception – books. You don't get to choose a format that is easy for you to read – you have to read the format the publisher selects.

This **"One Size Fits All"** paradigm **no longer** need apply to books.

At www.ReadHowYouWant.com, you can order a book just as you want it, so you can read it easily. You can choose nearly any format, and at a surprisingly low cost, because of a technical breakthrough that allows us to typeset an individual book automatically, print and bind the book and have it quickly sent directly to you.

You can have your book in **Talking Book** formats (**DAISY or MP3**) or as an **E-book** or in **Braille**.

We have developed totally new visual formats which we hope will help people with **dyslexia** and other print reading disabilities. Look at www.ReadHowYouWant.com for more information.